NOTEBOOK NO. ______________

AF371574

CONTINUATION FROM NOTEBOOK ____

____

NAME ____________________________________

DATE ISSUED ______________________________

INSTITUTION/COMPANY _____________________

____

____

DEPARTMENT ______________________________

EMAIL ___________________________________

ADDRESS _________________________________

____

____

PHONE ___________________________________

TABLE OF CONTENTS

TITLE   PROJECT

SIGNED   DATE

TITLE   PROJECT

SIGNED   DATE

TITLE   PROJECT

SIGNED   DATE

TITLE

PROJECT

SIGNED

DATE

TITLE   PROJECT

SIGNED   DATE

TITLE

PROJECT

SIGNED

DATE

TITLE   PROJECT

SIGNED   DATE

TITLE
PROJECT
SIGNED
DATE

TITLE   PROJECT

SIGNED   DATE

TITLE   PROJECT

SIGNED   DATE

TITLE   PROJECT

SIGNED   DATE

TITLE

PROJECT

SIGNED

DATE

TITLE

PROJECT

SIGNED

DATE

TITLE

PROJECT

SIGNED

DATE

TITLE   PROJECT

SIGNED   DATE

TITLE   PROJECT

SIGNED   DATE

TITLE   PROJECT

SIGNED   DATE

TITLE

PROJECT

SIGNED

DATE

TITLE

PROJECT

SIGNED

DATE

TITLE   PROJECT

SIGNED   DATE

TITLE

PROJECT

SIGNED

DATE

TITLE

PROJECT

SIGNED

DATE

TITLE

PROJECT

SIGNED

DATE

TITLE

PROJECT

SIGNED

DATE

TITLE   PROJECT

SIGNED   DATE

TITLE

PROJECT

SIGNED

DATE

TITLE   PROJECT

SIGNED   DATE

TITLE   PROJECT

SIGNED   DATE

TITLE

PROJECT

SIGNED

DATE

TITLE   PROJECT

SIGNED   DATE

TITLE   PROJECT

SIGNED   DATE

TITLE

PROJECT

SIGNED

DATE

TITLE   PROJECT

SIGNED   DATE

TITLE

PROJECT

SIGNED

DATE

TITLE
PROJECT
SIGNED
DATE

TITLE

PROJECT

SIGNED

DATE

TITLE
PROJECT
SIGNED
DATE

TITLE
PROJECT
SIGNED
DATE

TITLE
PROJECT
SIGNED
DATE

TITLE   PROJECT

SIGNED   DATE

TITLE

PROJECT

SIGNED

DATE

TITLE

PROJECT

SIGNED

DATE

TITLE
PROJECT
SIGNED
DATE

TITLE

PROJECT

SIGNED

DATE

TITLE

PROJECT

SIGNED

DATE

TITLE

PROJECT

SIGNED

DATE

TITLE   PROJECT

SIGNED   DATE

TITLE

PROJECT

SIGNED

DATE

TITLE

PROJECT

SIGNED

DATE

TITLE

PROJECT

SIGNED

DATE

TITLE
PROJECT
SIGNED
DATE

TITLE   PROJECT

SIGNED   DATE

TITLE   PROJECT

SIGNED   DATE

TITLE
PROJECT
SIGNED
DATE

Made in the USA
Monee, IL
07 July 2026

56551529R00059